Simply Sizzix

The Die-Cutter's Companion

Over 100 projects to inspire you, featuring award-winning Sizzix Personal Die-Cutting Sy...

by
Sandi Genovese
& Desirée Tanner
and
The Design Teams
of
Provo Craft
& Ellison

Published by
Sizzix LLC

Table of Contents

Credits

Provo Craft Design Team
Desirée Tanner
Christina Cole
Diane Garding
Sally Garrod
Jenny Jackson
Heather Lancaster
Cindy Schow
Shannon Wolz

Ellison Design Team
Sandi Genovese
Kim Fitzgerald
Cara Mariano

Layout by E.L. Smith, Lisa Cronin
and Jane Brenlin
Introductions Written by
Heather Hawkins and Sandi Genovese
Coordination by Susan Westerland
Copy Editing by Kay Elzea
Photography by J. Brenlin Design
Printing by Inland Color Graphics

Library of Congress Catalog Card Number
ISBN 0-9722380-0-X
First Edition - Printed in the United States of America

Introduction

WARNING: The projects shown in this book may cause the reader to squeal uncontrollably and break out into dance moves that haven't been seen since the 60's. So, close the windows and dust off your go-go boots. They are just that good.

If you currently own the award-winning* Sizzix Personal Die-Cutting System, you've probably thought many times, "How on Earth did I ever get anything done before Sizzix came along?" If you are one of the nine people who don't own a Sizzix Machine and Dies, your time has come. Sizzix has revolutionized home crafting by providing a perfect foundation for fabulous projects. It has opened up creative doors for scrapbookers and crafters of all abilities, and the wonderful array of projects in this book will provide even more inspiration.

While it is a "must-have" tool for scrapbookers, the versatility of the Sizzix System allows it to be used for a multitude of craft projects. The die-cutter can cut a wide range of materials – such as paper, cardstock, sheet magnet, self-adhesive rubber, shrink film, poly foam, fabric and felt – for creating personalized gifts, children's crafts, party decorations, cards, scrapbook enhancements, clothing and the list goes on. Patterns for the larger die-cut Cards, Boxes and Bags are provided in the back of the book.

This book has been divided into five chapters: Fun with Photos (scrapbook projects); Glorious Greetings (cards); Wrap It Up (gift wrap); Cut-It Make-It Wear-It (clothes and accessories); and Home & Play (décor and children's crafts). The projects are based on a few simple techniques. Once you've learned the techniques, you can easily carry them over to other projects. For example, say you fall in love with the Pig die-cut on the scrapbook layout on page 5. (Go ahead, take a look. How could you not fall in love with that little piglet?) That Pig can easily become a refrigerator magnet, a finger puppet or an appliqué on a little girl's dress simply by changing the materials.

Likewise, say you want to incorporate the wreath design on page 4 and 5 into one of your scrapbook pages, but it doesn't quite fit your theme. By choosing a different die design and changing the paper, you can adapt it to any theme! Of course you can simply copy the designs presented here, but it is so much more fun and rewarding to exercise the right side of your brain and come up with your own creative designs.

So, prepare to be inspired. And remember to close the windows. Enjoy!

*The Sizzix Personal Die-Cutting System won the 2002 Crafts/Craftrends Award of Excellence for "Best Scrapbooking Tool."

"Flower Power"

Getting Ready:

Sizzix Dies: Flower (Daisy #2), Flower (Rose),
Ladybug, Leaf Stem
Plain Paper: Canson
Letter Stickers: Making Memories
Punches: Fiskars, 1/16" & 1/8" hole punches
Pop Dots: All Night Media
Adhesive: Xyron

Quick Tip

Plates or bowls
(paper or ceramic) are
good guides for
creating perfectly
round rings.

Technique:

Create a ring out of paper that matches the background
paper. (Be sure that the ring is the right size for the
photo.) When the die-cuts are attached to this ring, the
wreath around the photo will be perfectly round. Any place
the ring is exposed will disappear because it is cut from
the same paper on which it rests.

*The same techniques can be used
to create a "hanging wreath" using
ribbon or string as an accent.*

Fun with Photos

"Animal Attraction"

Getting Ready:

Sizzix Dies: Bird, Frog, Giraffe, Pig, Sheep, Turtle
Plain Paper: Canson
Patterned Paper: Keeping Memories Alive
Adhesive: Xyron
Computer
Printer

For a wreath that is perfectly symmetrical, cut two of each animal. Be sure the patterned side of the paper is facing correctly before cutting.

Quick Tip

Please Welcome
Megan Rose Lee
8 pounds 4 ounces
20 1/2 inches long
Born 6:43 a.m.
4-22-2001

Technique:

Decorate a birth announcement to add to your scrapbook. Die-cut shapes from patterned papers to create a wreath that needs very little embellishment. The pattern in the paper is all the detail needed.

Fun with Photos

"Leo Got

Getting Ready:

Sizzix Dies: Bare Tree, Branch & Leaves,
Cloud #1, Cloud #2, Dog, Splats, Shadow Box Alphabet
Plain Paper & Cardstock: Provo Craft & Sizzix
Patterned Paper: Provo Craft
Chain: Teeny Weeny
Pop Dots: All Night Media
Glue Pen: Zig
Pen: Sakura
Adhesive: Xyron

Yech! What's that awful smell? It seems that Leo got up close & personal with a skunk! We tried shampoo, odor-off and even gave him a tomatoe sauce bath........

LEO GOT

Skunked"

Quick Tip

To elongate the tree trunk, cut where trunk meets photo and slide top of tree upward. Cut or draw woodgrain swirls onto the trunk.

...but he still smelled. It was truly the worst smell ever. He initiated our couch before it aded away. September '02

SKUNKED

Technique:

To add dimension to die-cuts, like the Dog and the Bare Tree, cut several in different colors. For the Dog, trim out the ear and leg and place Pop Dots between layers. Cut black paper for the nose and eyes. To give depth to the Bare Tree, use Clouds as leaves and adhere with Pop Dots.

Fun with Photos

7

"Some Bunny

Getting Ready:

Sizzix Dies: Basket, Bird, Bunny, Eggs, Grass, Scallop Oval Frame
Ellison Dies: 4 1/2" Petal Envelope, Grass Patch
Plain Paper: Canson
Patterned Paper: Provo Craft & Paper Patch
Stickers: Mrs. Grossman's
Punches: Fiskars, 1/8" & 1/16" hole punches
Pop Dots: All Night Media
Adhesive: Xyron
Ribbon

Quick Tip

To eliminate sticker waste on a Design Lines sticker, use the negative space as well as the positive space when decorating a card or scrapbook page.

Loves You"

Some Bunny Loves You

Technique:

Add movement and life to a scrapbook page by adding interactive elements such as cards or envelopes. Decorate the card with the negative parts of Design Lines stickers and keep it closed with coordinating ribbon.

"Free as a Bird"

Getting Ready:

Sizzix Dies: Cloud #1, Cloud #2, Hearts
Plain Paper & Cardstock: Provo Craft & Sizzix
Letter Stickers: Provo Craft
Sparkles: Art Accents & Shimmerz
Adhesive: Xyron
Scissors
Sand

Quick Tip

Match the shoreline on the page to the shoreline in the photos. Make seagulls from the top of Heart die-cuts.

Free as a Bird . . .

Technique:

To get the look of sand on a scrapbook page, use...sand! Mix a little sand with white sparkles to give the shoreline a realistic look.

"Joy"

Getting Ready:

Sizzix Die: Squares
Plain Cardstock: Provo Craft
Vellum: Paper Adventures
Alphabet Template: Provo Craft
Pen: Sakura
Font: Hugware
Adhesive: Xyron

Quick Tip

Type journaling in a fun font, leaving space for cut-out letters.

JOY

Elise has a great time at the cabin! That blanket she has over her head is DEALLy old! (This blanket is one of three that my mom got us when we were really young). Attila, Arpad and I all had those blankets growing up at the cabin... mine was orange and the boys' were blue. They're quite beat up now but still in use! Elise likes to hide under it and draw!

"Sometimes your joy is the source of your smile, but sometimes your smile is the source of your joy."
— Thich Nhat Hanh

july 2000

Technique:

Use vellum cut into Squares to create a geometric look. Embellish Squares with black pen for creative flourish. Overlapped corners give the page extra depth.

"Bottoms Up!"

Getting Ready:

Sizzix Dies: Hearts, Moon, Primitive Hearts,
Rectangle Frame, Stars
Ellison Dies: Gift Card & Insert
Plain Paper: Canson
Patterned Paper: Keeping Memories Alive
Vellum: Paper Adventures
Punch: Fiskars, Star
Pop-Dots: All Night Media
Adhesive: Xyron
Ribbon

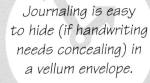

Quick Tip

Journaling is easy to hide (if handwriting needs concealing) in a vellum envelope.

Technique:

The use of multiple mats creates a special look that is also easy to make.
Select papers that coordinate, then vary the width of each mat. The Rectangle
Frame (trimmed) is perfect for matting the cropped photo within the yellow mat.
Position parts of the photo to overlap the Frame for a realistic look.

"Beautiful Things"

Getting Ready:

Sizzix Dies: Leaf Trio, Photo Corners
Plain Paper: Paper Adventures
Patterned Paper: Paper Patch
Vellum: Paper Adventures
Colored Pencils: Sanford
Fonts: Creating Keepsakes
Adhesive: Xyron
Ribbon
Scissors

Quick Tip

Avoid seeing double stick tape, glue or other adhesives through vellum by using decorative photo corners.

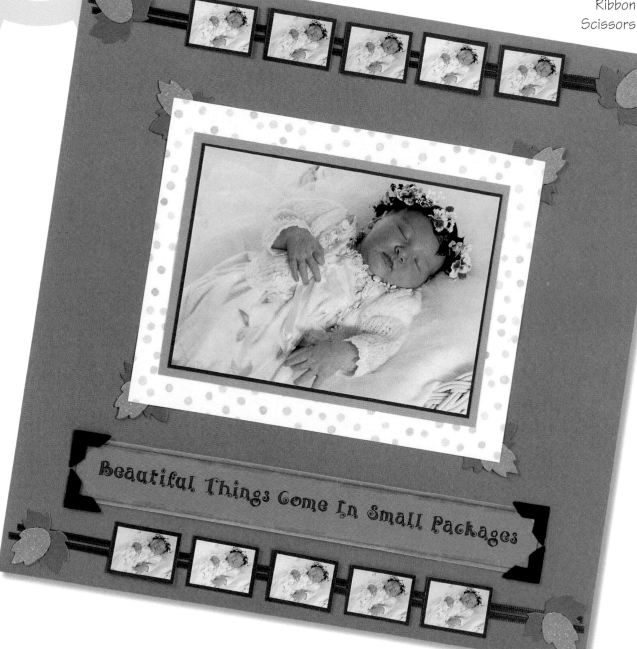

Beautiful Things Come In Small Packages

Technique:

Highlight a beautiful, enlarged photo in the center of the project by including repeated smaller images. Use the multiple image feature on color copiers to create wonderful mini photos.

Fun with Photos

"Ferry Good Day"

Getting Ready:

Sizzix Dies: Bitty Body, Bitty Boy Hair #1, Bitty Girl Hair #1,
Bitty Overalls, Bitty Shorts & Top, Bitty Swimsuit, Car,
Sailboat, Wave, Fun Serif Alphabet
Plain Paper & Cardstock: Provo Craft & Sizzix
Patterned Paper & Cardstock: Provo Craft & Sizzix
Vellum Sticker: Pathways
Letter Stickers: Provo Craft
Adhesive: Xyron
Cloth Covered Wire
Scissors

Quick Tip

Trim the mast and sail from the Sailboat die-cut to make a ferry or speedboat. Layer two Sailboats without the mast and sail to create the towed portion of the ferry.

Technique:

Add action to paper doll pages. Make the dolls sit by making a rounded cut at the knee. Adhere at waist level to the body.

"Sunshine"

Getting Ready:

Sizzix Dies: Fun Serif Alphabet
Plain Paper & Cardstock: Provo Craft
Patterned Paper: Provo Craft
Letter Stickers: Provo Craft
Adhesive: Xyron
Scissors

Quick Tip

To create rays of sunlight, lay strips of yellow paper from the sun to the edges of the paper before placing the photo and mats on the page. This will ensure a natural fan out of sun rays.

Our Sunshine

Summer 2001

Technique:

Create a multi-colored mat around a photo by first placing the photo on black cardstock 1/2" larger than the photo. Place scraps of different colored papers on another black mat that's about 3/4" larger than the first. Trim so sides are even and finish off with a long strip of black paper around the outside edge.

"Photo Wheel"

Getting Ready:

Sizzix Dies: Apple, Balloon #1, Dragonfly, Filmstrip
Ellison Dies: Clipboard, Flash Card Wheels,
Lollipop Alphabet
Plain Paper: Canson
Patterned Paper: Paper Adventures
Stickers: Mrs. Grossman's
Star Brad: American Pin
Adhesive: Xyron

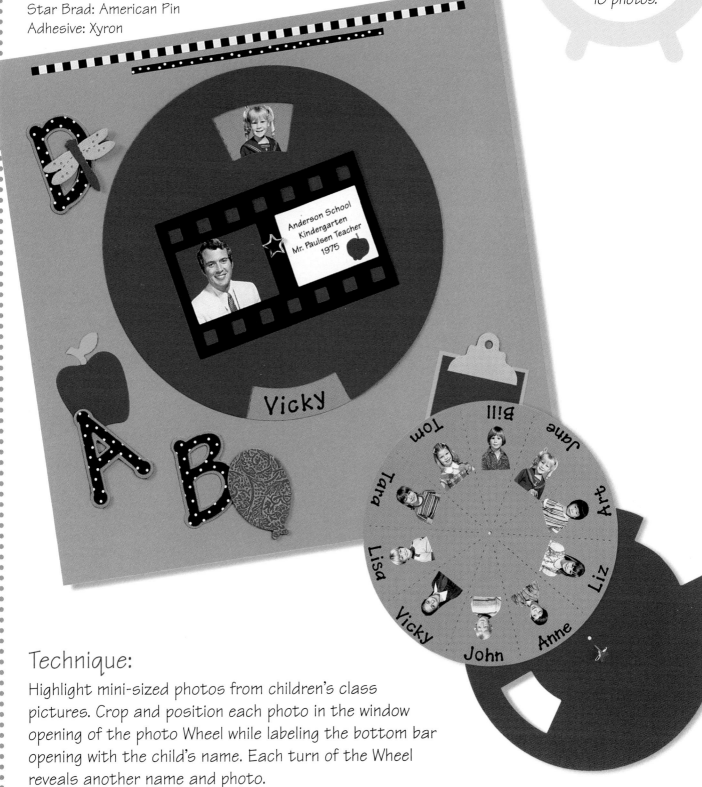

Quick Tip

Lay the Top Wheel over the Bottom Wheel when positioning each photo and name to insure perfect placement. Each Wheel holds 10 photos.

Anderson School
Kindergarten
Mr. Paulsen Teacher
1975

Vicky

Tom · Bill · Jane · Art · Liz · Anne · John · Vicky · Lisa · Tara

Technique:

Highlight mini-sized photos from children's class pictures. Crop and position each photo in the window opening of the photo Wheel while labeling the bottom bar opening with the child's name. Each turn of the Wheel reveals another name and photo.

"Spring"

Getting Ready:

Sizzix Dies: Flower #1, Leaf Stem, Fun Serif Alphabet
Ellison Dies: Flower Pot
Plain Paper: Canson
Patterned Paper: Paper Patch
Rubber Stamp: JudiKins
Stamp Pad: Versa Mark by Tsukineko
Punch: Fiskars, Flower
Pop Dots: All Night Media
Adhesive: Xyron
Craft Knife

Quick Tip

To eliminate busy backgrounds in a photograph cut around the subjects and place them in a new background or scene.

Serina in the garden. June, 2001

Technique:

Create a shadow box with a title. Die-cut the alphabet and place the letters on the page using removable tape. With a craft knife cut around the letters creating the negative space in your page. Add Pop-Dots to the back of the whole page and attach to a larger colored page to create the dimension for the shadow box letters.

"Bunny

Getting Ready:

Sizzix Dies: Primitive Hearts, Tags
Plain Paper: Provo Craft
Patterned Paper: Provo Craft
Velveteen Paper: Wintech
Vellum: Paper Adventures
Cutting Mat: Provo Craft
Chalks: Craft-T Products
Embroidery Floss: DMC
Eyelets: Lasting Impressions
Eyelet setting tool
Font: Hugware
Glue Pen: Zig
Adhesive: Xyron
Hammer
Scissors
Button

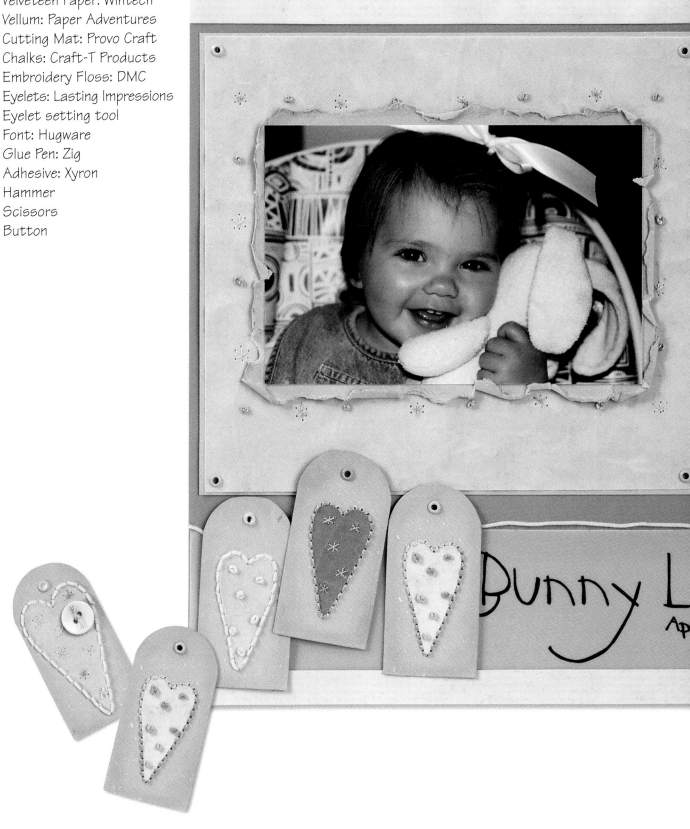

Love"

Quick Tip

Poke small holes in the page to show where the embroidery will go.

Aiyanna has attached herself to this cute pink bunny. Bunny has been on many trips with us and has been washed several times— that is love! Aiyanna is such a girl, loving all of Ian's discarded dollies.

Technique:

Add dimension to a scrapbook page by embroidering embellishments on photo mats and die-cuts. Starbursts and french knots are simple but elegant additions that can be carried through all the elements on a page.

"Our Little Critter"

Getting Ready:

Sizzix Dies: Bird, Bunny, Cat, Dog,
Green Tree, Ladybug, Pig, Turtle
Plain Paper: Canson
Stickers: Mrs. Grossman's
Little Letters: Making Memories
Rubber Stamp: JudiKins
Stamp Pad: Versa Mark by Tsukineko
Pop Dots: All Night Media
Adhesive: Xyron

Quick Tip

Use removable labeling tape on the page as a guide when using letter stickers to eliminate crooked titles.

OUR LITTLE CRITTER

Technique:

Create custom patterned paper with a rubber stamp for the photo background. Using a watermark stamp pad and a background stamp, repeat the pattern to cover the whole sheet of paper. Cut to size and mat it on a contrasting color of paper.

Fun with Photos

"Eggcellent"

Getting Ready:

Sizzix Dies: Bird, Eggs
Plain Paper: Canson
Patterned Paper: Paper Patch
Stickers: Mrs. Grossman's
Rubber Stamp: JudiKins
Stamp Pad: Versa Mark by Tsukineko
Punches: Fiskars, 1/16" & 1/8" hole punches
Pen: Sakura
Pop Dots: All Night Media
Adhesive: Xyron

Quick Tip

To create polka-dot Eggs, cut two Eggs in contrasting colors of paper. Punch multiple holes in the top Egg and fasten it to the solid Egg underneath. The color of the bottom Egg will show through.

Technique:

When creating a border with multiple die-cuts, attach them to a strip that matches the background paper. Then fasten the decorated strip to the page. This allows for easy repositioning and makes it easy to keep the border design straight and parallel to the bottom of the page.

"Window Box"

Getting Ready:

Sizzix Dies: Filmstrip (Squares), Squares
Plain Paper: Canson
Stickers: Mrs. Grossman's
Pop Dots: All Night Media
Adhesive: Xyron

Quick Tip

To miter the corners of the window, cut strips long enough to overlap. Attach the strips in the middle and cut diagonally through the overlap before adhering the corners.

Technique:

Enlarge a photo, then cut it in half sideways and longways. Pull each section slightly apart when attaching it to the background to create a window-like effect. Use the drop-out squares from the Filmstrip die-cut and additional strips of paper to further enhance the look of a window with larger Squares filled with raised sticker flowers.

"Girls, Girls, Girls"

Getting Ready:

Sizzix Dies: Doll Body, Doll Dresses, Doll Girl Hair #1
Plain Paper: Canson & Paper Adventures
Patterned Paper: Paper Patch & Provo Craft
Rubber Stamps: Provo Craft & JudiKins
Stamp Pad: Versa Mark by Tsukineko
Colored Pencils: Sanford
Pens: Zig
Adhesive: Xyron

Technique:

Create original patterned paper by stamping on plain colored paper or cardstock with a background rubber stamp and a watermark stamp pad. Rubber stamp the paper before die-cutting.

"Bailey"

Getting Ready:

Sizzix Dies: Flower (Daisy #2), Leaf Trio, Squares
Plain Cardstock: Provo Craft & Sizzix
Patterned Paper & Cardstock: Provo Craft
Pen: Sakura
Chalk: Craft-T Products
Eyelets: Doodlebug
Glue Pen: Zig
Paintbrush
Ribbon
Scissors

Technique:

Create peek-a-boo squares by folding a rectangle in half and cutting with the fold intact. Use each panel as a separate page for journaling.

"May"

Getting Ready:

Sizzix Dies: Butterfly #1, Flower (Daisy #2), Photo Corners
Paper & Cardstock: Provo Craft & Sizzix
Pen & Watercolor Pencils: Sakura
Woven Twine: Pulsar
Pop Dots: All Night Media
Glue Pen: Zig
Adhesive: Xyron
Wire Tweezers & Cutters
Colored Wire
Buttons
Scissors
Raffia

Quick Tip

Unroll the raffia so it's flat and run it through the Xyron. Place on white cardstock and die-cut into Butterflies and Daisies.

In Provo, 2001

Technique:

Use watercolor pencils that coordinate with paper colors to add soft dimension to scrapbook pages. Dip the lead into water and draw directly on the paper. Watercolor pencils can also be used dry. Draw on the paper and then go over the drawing with a moist brush to create a painted watercolor look without the mess of paint.

Fun with Photos

"Love"

Getting Ready:

Sizzix Dies: Hearts, Rectangle Frame, Squares,
Shadow Box Alphabet
Plain Paper: Canson & Tru-Ray
Patterned Paper: Paper Patch & Provo Craft
Paint Pen: Marvy/Uchida
Eyelets: Lasting Impressions
Adhesive: Xyron

Quick Tip

Check the fine arts
section of craft stores
where paints and brushes
are displayed to find
large sheets of paper
in a rainbow range
of colors.

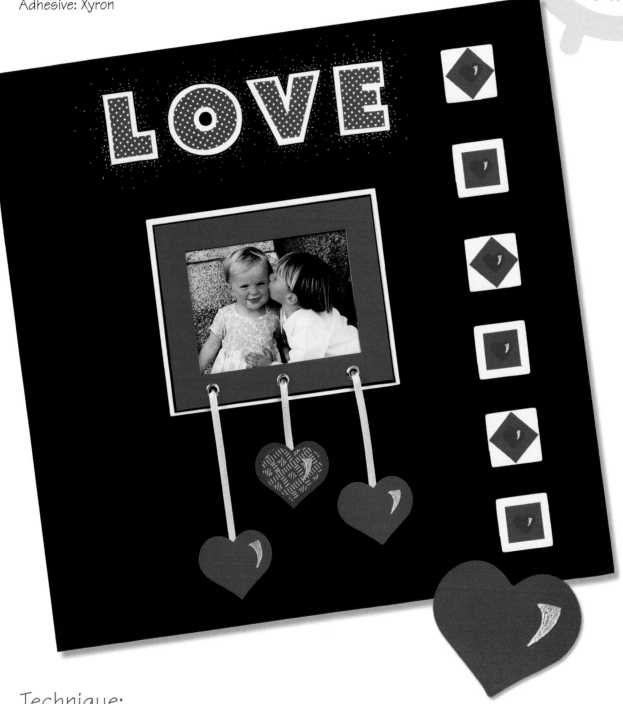

Technique:

Ribbon, wire, thread or raffia can be threaded through decorative eyelets that have
been attached to the project. Die-cuts, tags, buttons, and other items can now
dangle from projects to add a lot of fun and movement.

Getting Ready:

Sizzix Die: Tags
Ellison Dies: Acrylic Rectangle, Rectangle
Plain Paper: Canson
Patterned Paper: Paper Patch
Vellum: Paper Adventures
Stickers: Mrs. Grossman's
Punch: Fiskars
Brads: American Pin & Fastener
Font: Creating Keepsakes
Pop Dots: All Night Media
Pen: Zig
Adhesive: Xyron

Technique:

Rather than writing directly on a project, add tags for a unique way to journal or add a personal message. Cut slits the width of the tag with a craft knife in a strip of vellum so they can be easily added or removed.

"Thankful"

Getting Ready:

Sizzix Dies: Bow, Squares, Tiny Leaf #1,
Tiny Leaf #3, Shadow Box Alphabet
Plain Cardstock: Sizzix
Patterned Paper: Scrap-Ease
Cord: Pulsar Paper Products
Pop Dots: All Night Media
Adhesive: Xyron & Hermafix

Quick Tip

Use Square die-cuts folded over the corner of a photo to make triangle photo corners.

Technique:

To create "crunchy" looking autumn leaves, soak cardstock in water for a few minutes, crinkle it and let it dry overnight. Once it's dry, cut out leaves.

"Sundance"

Getting Ready:

Sizzix Dies: Mountain Border, Photo Corners,
Road & Wavy Border, Sand or Snow Mound,
Fun Serif Alphabet
Plain Cardstock: Provo Craft
Patterned Paper: Doodlebug
Vellum: Paper Adventures
Sticker: Provo Craft
Pop Dots: All Night Media
Adhesive: Xyron
Twine

Technique:

To create gentle waves of mountains, carefully tear paper into strips. Carry this natural look through the page by using torn paper for photo mats and titles.

"Hope"

Getting Ready:

Sizzix Dies: Flower #1, Leaf Stem, Fun Serif Alphabet
Ellison Dies: 5-Page Mini Book Card
Plain Paper: Anna Griffin, Paper Adventures & Canson
Patterned Paper: Anna Griffin
Rubber Stamp: JudiKins
Stamp Pad: Tsukineko
Adhesive: Xyron
Scissors

Quick Tip

Enlarge photos on a color copier by using the following formula: 100% x desired size ÷ original size = enlargement %. Example: 100% x 9" = 900; 900 ÷ 6" = 150%.

Technique:

When a die-cut calls for shading to create detail and dimension, use Dauber Duos stamp pads to create a wonderfully subtle effect. Start with the lighter shade and lightly feather in the darker color to create a sense of depth.

"Bride In Training"

Getting Ready:

Sizzix Dies: Squares, Tags
Ellison Die: Plain Greeting Card
Plain Paper: Canson
Patterned Paper: Keeping Memories Alive
Stickers: Mrs. Grossman's
Pen: Sakura
Pop Dots: All Night Media
Adhesive: Xyron
Ribbon

Quick Tip

Use coordinating stickers that match the scrapbook page. Add adhesive to attach the ribbon to the journaling tags.

Technique:

Attach pockets to a card or scrapbook page to provide a spot to place finished tags. To create the pocket, cut two rectangles making sure one is slightly larger than the other and layer them together. Attach double-sided tape along the sides and bottom of the matted rectangles, but leave the top open for tag insertion. Attach the finished rectangles to a card. Place the finished tags in the pockets.

Fun with Photos

31

"Snow Bunnies"

Getting Ready:

Sizzix Dies: Grass, Snowman, Wood Sign
Plain Cardstock: Provo Craft & Sizzix
Patterned Paper: Provo Craft
Letter Stickers: Provo Craft
Watercolor Pencil: Sakura
Eyelets: Doodlebug
Pop Dots: All Night Media
Adhesive: Xyron

Add ears and tail to the Snowman to transform him into a snow bunny, snow cat or snow dog.

SNOW bunnies

Bailey made a snowbunny in our front yard- complete with a celery nose! December '01

Technique:

Use eyelets to embellish die-cuts or to adhere photos to a scrapbook page for a fun effect.

"Santa's Lap"

Getting Ready:

Sizzix Dies: Santa Head, Squares, Shadow Box Alphabet
Plain Cardstock: Provo Craft & Sizzix
Patterned Paper: Provo Craft
Letter Stickers: Provo Craft
Adhesive: Xyron
Snow in a Bottle
Paint Brush
Ribbon
Sock

Quick Tip

Run ribbon through the Xyron and lay it side by side on white cardstock before die-cutting letters.

SANTA'S LAP.

"is a warm and fuzzy place to be"

USC Christmas Recital

Bailey · 2001

Technique:

Die-cut the inside of a sock to add a fuzzy look to Santa and the shadow of the Shadow Box Alphabet. Snow in a Bottle, brushed on die-cuts, adds extra fuzziness for a fun holiday page.

Fun with Photos

33

"Flower Slider"

Getting Ready:

Sizzix Dies: Dragonfly, Flower (Daisy #2),
Ladybug, Leaf Stem
Ellison Dies: Flower (Daisy #2), Slider Card Set
Plain Paper: Canson
Punches: Fiskars, 1/16" & 1/8" hole punches
Pop Dots: All Night Media
Adhesive: Xyron

"More Flower Slider"

Quick Tip

Tell the viewer to lift first on the card front, and label "pull" on the slider tab to make it easy to see the photos in the order you intended.

PULL

PULL

Technique:

Chronicle kids at three different times in their lives on an interactive scrapbook page that holds a Slider Card. Feature the youngest photo on the card front, the middle age photo under the card flap (visible when the front cover is lifted) and the final photo behind the slider portion of the card, visible when the slider is pulled.

Fun with Photos

"My Journal"

Getting Ready:

Sizzix Dies: Dragonfly, Squares, Fun Serif Alphabet
Plain Cardstock: Sizzix
Patterned Paper: Provo Craft
Letter Stickers: Provo Craft
Adhesive: Xyron
Album or Journal

Quick Tip

Use clear shelf paper to keep the cover in perfect condition.

Technique:

Ghosted letters provide a fun and easy way to title your journal. Cut letters out of two papers that coordinate with the cover background paper and stack. This gives the title enough contrast from the background.

"Summer Vacation"

Getting Ready:

Sizzix Dies: Bubbles, Fish, Squares
Plain Paper: Canson
Patterned Paper: Provo Craft
Vellum: Paper Adventures
Punches: Fiskars, 1/16" & 1/8" hole punches
Pens: Sakura
Font: Creating Keepsakes
Pop Dots: All Night Media
Adhesive: Xyron
Album or Journal

Quick Tip

Create custom colored titles by using fill-in fonts and colored pencils or marking pens.

Summer Vacation 2001

Technique:

Change the look of any journal by adding a new color and vellum overlay to the cover. Detach the journal cover by removing it from the journal rings. Cut paper to size and attach to old cover. Using a hole punch turn the cover over and punch new holes by using the original cover as the guide. Do the same technique with vellum for a custom overlay.

Fun with Photos

"Boo-T-Full Journal"

Getting Ready:

Sizzix Dies: Ghosts, Pumpkin, Tags
Ellison Die: 3 3/4" Square (Acrylic)
Plain Paper: Canson
Patterned Paper: Paper Patch
Pens: Sakura
Letter Stickers: Making Memories
Adhesive: Xyron
Punch: Fiskars
Ribbon

Quick Tip

If the die-cut shapes have been Xyroned, place an empty sticker backing sheet behind the window opening when attaching the die-cuts. This will prevent them from sticking to the inside page where they overlap the window opening.

BOO!

BOO!

Technique:

Create personalized journals using large sheets of paper, folded in half with a square window opening in the cover. The square can be die-cut in the Ellison Machine, hand cut with a Coluzzle template and swivel knife, or cut with a craft knife and ruler. Die-cuts can overlap the window opening, cutting each design in matching pairs, making sure the pair are mirror images.

"Mini Memory Book"

Getting Ready:

Sizzix Dies: Flower (Daisy #2), Ladybug, Tags
Ellison Die: 2 3/4" Square (Acrylic)
Paper: Canson
Punches: Fiskars, 1/16" & 1/8" hole punches
Pen: Sakura
Pop Dots: All Night Media
Adhesive: Xyron
Ribbon

Quick Tip

Be sure to cut the inside pages smaller than the cover to avoid any pages peeking outside the cover when the book is closed.

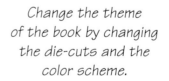

Change the theme of the book by changing the die-cuts and the color scheme.

Technique:

Label each journal with a tag that is threaded through the ribbon before tying into a bow to hold the album together. The tag can be left to hang loose or can be held in one place with an adhesive dot.

Fun with Photos

39

"Jason and Shaya

Getting Ready:

Sizzix Dies: Jelly Frame, Rectangle Frame,
Scallop Frame, ZigZag Frame
Plain Paper: Canson
Patterned Paper: Paper Adventures
Letter & Number Stickers: Making Memories
Ribbon: Offray
Adhesive: Xyron

Technique:

Sandwich ribbon between matching pairs of picture
frames, leaving a little ribbon exposed in between each
frame to allow the completed strip of frames to be accordion folded and extra
ribbon on the end to tie it closed.

Ribbon Journal"

Quick Tip

The inside and outside mats in each frame are Rectangle Frame die-cuts. The inside mat is trimmed 1/4" from each side and layered on top of the untrimmed frame, with the decorative frame placed on top.

Front

Back

Because the ribbon is sandwiched between each pair of frames, the ribbon journal is two-sided: Jason's photos on one side and Shaya's photos on the flip side.

Glorious Greetings

Convert your favorite scrapbook tools into card-making tools to create one-of-a-kind greeting cards, stationery and note cards. Whether you're thanking, inviting or just saying hello, make it totally personal by making it yourself. Dive into card making and kiss those store bought cards goodbye!

"Happy Birthday"

Getting Ready:

Sizzix Die: Gifts
Ellison Dies: Window Card #7,
A6 Envelope, A6 Envelope Liner
Plain Paper: Canson
Patterned Paper: Paper Patch
Stickers: Mrs. Grossman's
Paint Pen: Marvy/Uchida
Pop-Dots: All Night Media
Adhesive: Xyron

Quick Tip

Use the computer to create quick titles in any size or font. Print multiples on one page to create several cards or save them for future use.

Happy Birthday

Technique:

Using a metallic silver paint pen, add sparkle around a border or die-cut. Start by adding small dots repeatedly close together near the border. The farther from the border the more spaced out the dots should be.

"Happy

Getting Ready:

Sizzix Dies: Jelly Frame, Party Favor,
Rectangle Frame
Ellison Die: Photo Envelope #1
Plain Paper: Canson
Patterned Paper: Paper Patch
Adhesives: Xyron & Double Stick Tape
Letter Stickers: Making Memories
Pop Dots: All Night Media

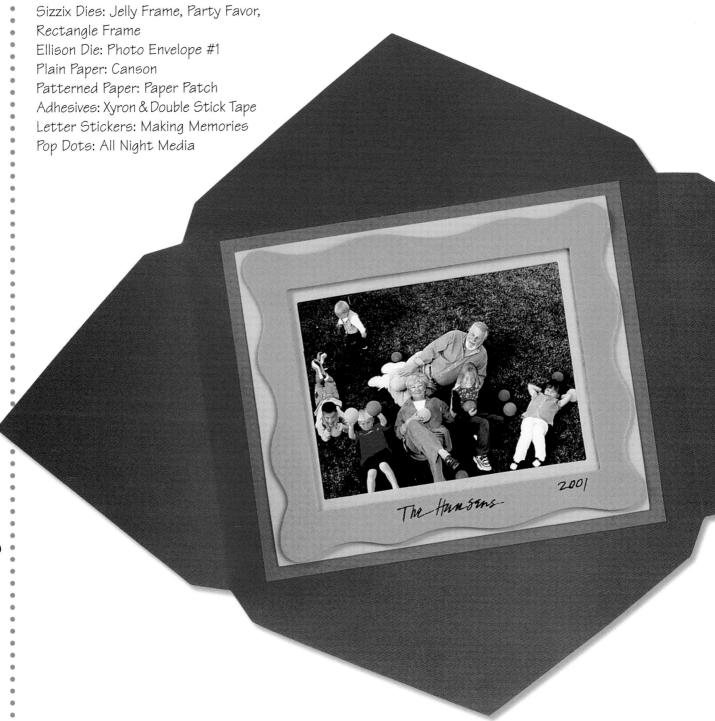

The Hansens   2001

Technique:

Create a removable wrap to keep a finished envelope closed. Layer two strips of contrasting colored paper and wrap it around the envelope. Keep it loose so it can easily slide on and off and attach the two ends on the back using double stick tape. Hide the seams by writing the address information on a separate rectangle and attaching it to the wrapper.

New Year"

Quick Tip

Save time and money when making color copies of a picture that will be used several times by pressing the repeat function on the copy machine. This provides up to four pictures on one sheet.

Happy New Year

John & Sarah James
753 Elm St.
Laguna Beach, Ca. 97153

LOVE
USA 34

Glorious Greetings

45

"Easter"

Getting Ready:

Sizzix Dies: Bunny, Hearts, Road & Wavy Border
Plain Cardstock: Provo Craft
Patterned Paper: Provo Craft
Self Adhesive Felt: Ellison
Shape Templates: Provo Craft
Swivel Knife & Cutting Mat: Provo Craft
Glue Pen: Zig
Scissors

Quick Tip

Place a small piece of blue or green paper under the Bunny die-cut to show through the eye.

Technique:

Use shape templates, such as the Coluzzle Cutting System to create a large oval shape to use as a card. Fold a piece of cardstock and lay the template on the paper past the fold. Cut through both layers of the folded paper. This will leave one side uncut to be the spine of the card.

"Dragonfly Dangle"

Getting Ready:
Sizzix Dies: Dragonfly, Ivy Border
Ellison Dies: Accordion w/Cut Out, Card Cover A6
Plain Paper: Canson
Stickers: Mrs. Grossman's
Rubber Stamp: JudiKins
Watermark Stamp Pad: Tsukineko
Adhesives: Xyron & Double Stick Tape
Pen: Marvy/Uchida
Thread

Quick Tip

Use the same colored thread as the background to give the illusion that the dangling element is "floating".

Technique:

Many die-cut cards have fold lines that allow the cards to become three-dimensional when opened. By attaching only the two end panels to the card cover, an area is created that is perfect for hanging pairs of die-cuts with thread sandwiched in between the matching die-cuts.

Glorious Greetings

47

"Double Heart"

Getting Ready:

Sizzix Die: Primitive Hearts
Plain Paper & Cardstock: Provo Craft
Patterned Paper: Provo Craft
Vellum: Paper Adventures
Punch: Fiskars
Pop Dots: All Night Media
Adhesive: Xyron
Glue Pen: Zig
Wire

Secure the wire
with dots of glue
beneath the hearts
so the glue
won't show.

Technique:

Embellish cards with decorative wire to accentuate the pattern in the paper used on the front of the card.

"Santa Card"

Getting Ready:

Sizzix Dies: Butterfly #3, Santa Head
Plain Cardstock: Provo Craft & Sizzix
Patterned Paper: Sizzix
Letter Stickers: Provo Craft
Chalk: Craft-T Products
Adhesive: Xyron
Twine

Quick Tip

Use the drop out part of the Butterfly die-cut for the small hearts on this card.

Merry Christmas

The Tanners

The Tanners

Technique:

When folding the card, make the cover shorter than the back so the die-cut on the front can hang below the bottom of the card front. The bottom of the die-cut will show when the card is opened.

Glorious Greetings

49

" I Be "Leaf" in You"

Getting Ready:

Sizzix Dies: Leaf #1, Leaf #2, Leaf #3
Ellison Dies: Card Cover A6, Multiple Pop-Up Card #1,
Plain Paper: Canson & Paper Adventures
Stickers: Mrs. Grossman's
Punches: Fiskars, 1/16", 1/8" & 1/4" hole punches
Pens: Pentel
Adhesive: Xyron

Quick Tip

If the die-cuts have been Xyroned, use powder and a small paint brush to remove the sticky on the leaf back where adhesive may be exposed on your project.

Technique:

Create leaf "veins" with gel or milky pens. Use complimentary shades of the same color on your leaves for added detail and dimension.

"Happy Halloween"

Getting Ready:

Sizzix Dies: Ghosts, Grass, Pumpkin
Ellison Die: Diorama Arch Card
Plain Paper: Canson
Sticker Letters: Making Memories
Pen: Marvy
Pop Dots: All Night Media
Adhesive: Xyron
Thread

Quick Tip

Hang shapes by die-cutting matching pairs and then sandwiching a piece of thread between the two shapes. Use thread that is the same color as the background so it disappears.

Technique:

Make any die-cut shape stand up by cutting it on the fold. First fold paper in half. Lay the folded edge of the paper just inside the blade of the die at the bottom of the design. Be sure the cutting blade extends below the edge of the fold. Then cut in the machine. This will create two shapes that are hinged together at the bottom. Tape one layer of the cut out to the card, and stand the other layer at a 90 degree angle.

Glorious Greetings

"Little Lady"

Getting Ready:

Sizzix Dies: Ladybug, Rectangle #1
Plain Cardstock: Provo Craft
Patterned Paper: Sizzix
Letter Stickers: Provo Craft
Eyelets: Doodlebug
Eyelet Setting Tool
Wire: Artistic Wire
Wire Cutters & Pliers
Mesh: Avant Card
Pen: Sakura
Fusible Bonding: Poly-Fil
Glue Pen: Zig
Adhesive: Xyron
Button
Hammer
Scissors

Quick Tip

Secure
the bow over the
heart button with a
dot of glue.

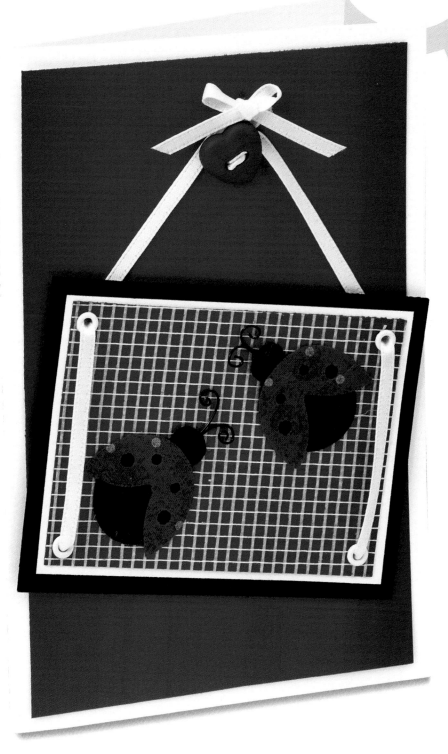

Glorious Greetings

Technique:

Because glues show on paper, use fusible bonding for materials like mesh or see-through ribbon.

"Patriotic"

Getting Ready:

Sizzix Dies: Primitive Stars, Squares, Star Border
Plain Cardstock: Sizzix
Patterned Paper & Cardstock: Provo Craft
Shape Template: Provo Craft
Swivel Knife & Cutting Mat: Provo Craft
Adhesive: Xyron
Foam: Westrim

Quick Tip

Use a foam Star as an envelope seal.

Technique:

Cut Primitive Stars from Fun Foam to create a 3-dimensional card. Foam is strong enough to hang off the card without worry of the shape getting bent or folded.

Wrap It Up

Gift wrap, gift tags, boxes and bags are all quick and easy with Sizzix. Cut sponge for original sponge painted wrapping paper. Or make an ordinary gift bag extraordinary with a few die-cuts and a little ingenuity.

"Flower Sponge Wrap"

Getting Ready:

Sizzix Dies: Branch & Leaves,
Flower (Daisy #2), Swirls
Pop-Up Sponge: Ellison
Butcher Paper
Paint: Delta
Ribbon: Offray
Paint tray
Water

Quick Tip

Wrap the present first to insure shapes will not get folded down and not show.

Technique:

Die-cut compressed Pop-Up Sponge then wet the sponge shape. Dip the sponge in paint and blot before pressing onto the paper. Sponge Flowers, let dry, then sponge Leaves.

Wrap It Up

"Decorative Boxes"

Getting Ready:

Sizzix Dies: Bat, Flower #1, Holly & Berries,
Leaf Stem, Squares
Ellison Die: Bag #1
Punches: Fiskars, 1/16" & 1/8" hole punches
Plain Paper: Canson
Adhesives: Xyron & Double Stick Tape
Tissue Paper: DMD Industries
Pen: Marvy/Uchida
Pop Dots: All Night Media

Quick Tip

Make double-sided paper to die-cut your favorite bag and create a stronger base for heavier objects.

Technique:

To add decorative detail to a die-cut, first cut the shape out of two colors, making sure one color is darker than the other. Draw in the design with pencil on the lighter shape. With craft scissors, cut away the design and layer onto the darker shape. This allows the detail to show up.

"Christmas Gifts"

Getting Ready:

Sizzix Dies: Holly & Berries, Ornament
Ellison Die: Box #9 Set
Plain Paper: Canson, Tru-Ray
Patterned Paper: Paper Patch
Rubber Stamp: Rubber Stampede
Adhesives: Xyron & Double Stick Tape
Stamp Pad: ColorBox
Pen: Marvy/Uchida
Eyelets: Impress
Ribbon

Quick Tip

As a general rule of thumb, elements such as die-cut Ornaments or Holly look better in odd numbers.

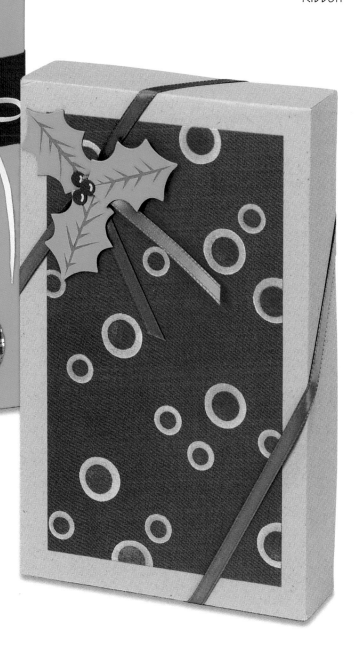

Technique:

After working hard to arrange and put together a project, give those hanging elements extra support by adding eyelets. They will give punched holes a finished look and help prevent the ribbon or paper from tearing through the holes.

Wrap It Up

57

"Autumn Gift Bags"

Getting Ready:

Sizzix Dies: Flower (Daisy #2), Leaf #1
Plain Cardstock: Provo Craft
Patterned Paper: Provo Craft
Tissue Paper: DMD Industries
Bags: DMD Industries
Punch: Fiskers, 1/8" hole punch
Ribbon: Offray
Mesh: Avant Card
Adhesive: Xyron
Twine

Quick Tip

Punch two small holes in the center of the large daisy and thread the ribbon through, just like you would on a button.

Technique:

Add die-cuts in warm hues, like yellows and browns to natural colored gift bags for an autumn feeling.

"Heart Gift Bags"

Getting Ready:

Sizzix Dies: Hearts, Primitive Hearts, Squares, Tags
Plain Paper & Cardstock: Provo Craft
Patterned Paper: Provo Craft
Corrugated Paper: DMD Industries
Bags: DMD Industries
Tissue: DMD Industries
Eyelets: Doodlebug
Mesh: Avant Card
Ribbon: Offray
Scissors
Twine

Quick Tip

Secure items such as eyelets or buttons to the die-cuts before placing on the bag.

Technique:

Use Squares or Rectangles as a mat for other die-cut shapes. Carry the design through to the gift tag.

Wrap It Up

"Clearly Flowers Bag"

Getting Ready:

Sizzix Dies: Flower (Daisy #2), Tags
Plain Cardstock: Sizzix
Patterned Paper: Provo Craft
Clear Gift Bag: DMD Industries
Tissue Paper: DMD Industries
Pop Dots: All Night Media
Pen: Zig
Glue Pen: Zig
Ribbon

Using a dot of glue on the top of each daisy will keep them from getting crinkled.

To Aiyanna

Technique:

Adhere die-cut Daisies on the inside of a clear gift bag for a great spring gift.
Top it off with a homemade gift tag from coordinating paper.

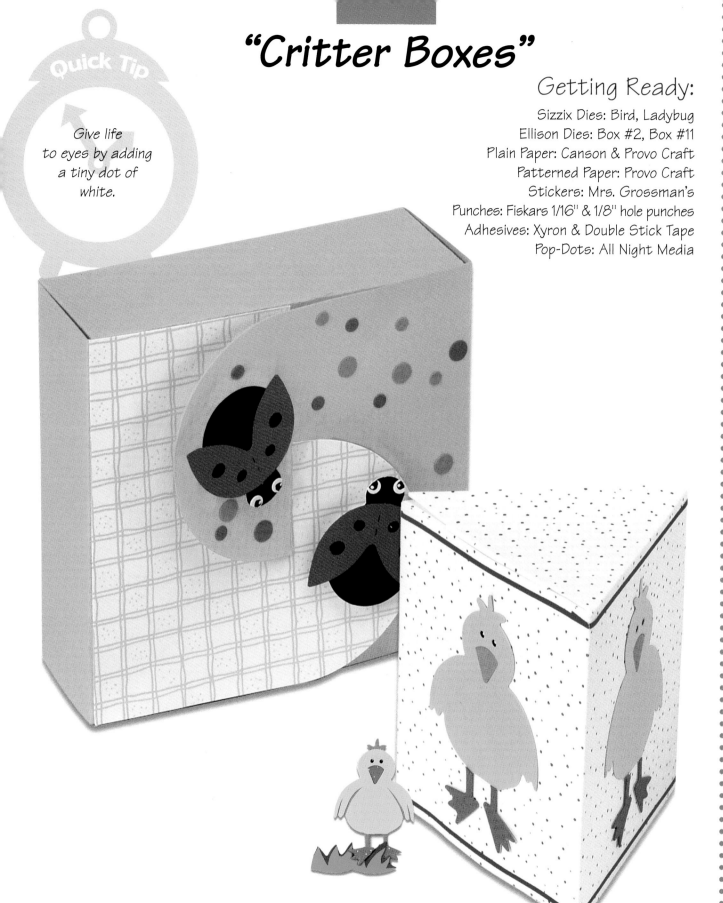

"Critter Boxes"

Getting Ready:

Sizzix Dies: Bird, Ladybug
Ellison Dies: Box #2, Box #11
Plain Paper: Canson & Provo Craft
Patterned Paper: Provo Craft
Stickers: Mrs. Grossman's
Punches: Fiskars 1/16" & 1/8" hole punches
Adhesives: Xyron & Double Stick Tape
Pop-Dots: All Night Media

Quick Tip

Give life to eyes by adding a tiny dot of white.

Technique:

Add embellished elements to the box while it is still flat but add sticker strips after assembly to avoid buckled or pulled corners.

Cut-It Make-It Wear-It

Wear it on your feet, around your neck or over your shoulder. Put your own special mark on tote bags, jewelry, sandals and everything in between. Die-cut foam, plastic and fabric to personalize your or your kid's wardrobe. Even die-cut rubber to create rubber stamped clothing that will be cherished forever.

"Flower Canvas Tote"

Getting Ready:

Sizzix Dies: Circles, Leaf Stem
Fabric Paint: Delta & Rubber Stampede
Clear Stamp Mounts: Ellison
Self-Adhesive Rubber: Ellison
Plain Canvas Tote Bag: Michaels
Paintbrush

Quick Tip

Empty film canisters (with the lid attached) make great stamp mounts for smaller die-cut shapes.

Technique:

Die-cut shapes from rubber and attach to stamp mounts to create personalized rubber stamps. Apply paint to rubber stamp with a paint brush or sponge applicator. Colors can be blended right on the rubber surface and need only be stamped once for shading on the design.

Cut-It Make-It Wear-It

63

"Bouncing Baby Boy"

Getting Ready:

Sizzix Dies: Basketball, Fun Serif Alphabet
Creeper: Pampers
Socks: Target
Pop-Up Sponge: Ellison
Paint: Delta
Embroidery Floss: Anchor
Fusible Bonding: Poly-Fil
Invisible Quilting Marker
Fabric
Needle
Scissors

Technique:

Die-cut Pop-Up Sponge. Wet the sponge shape and wring as much water as possible out of the sponge before applying paint. Use the sponge to paint the shape onto the creeper and socks. Die-cut letters from fabric and use fusible bonding to adhere to the creeper.

"Ladybug Creeper"

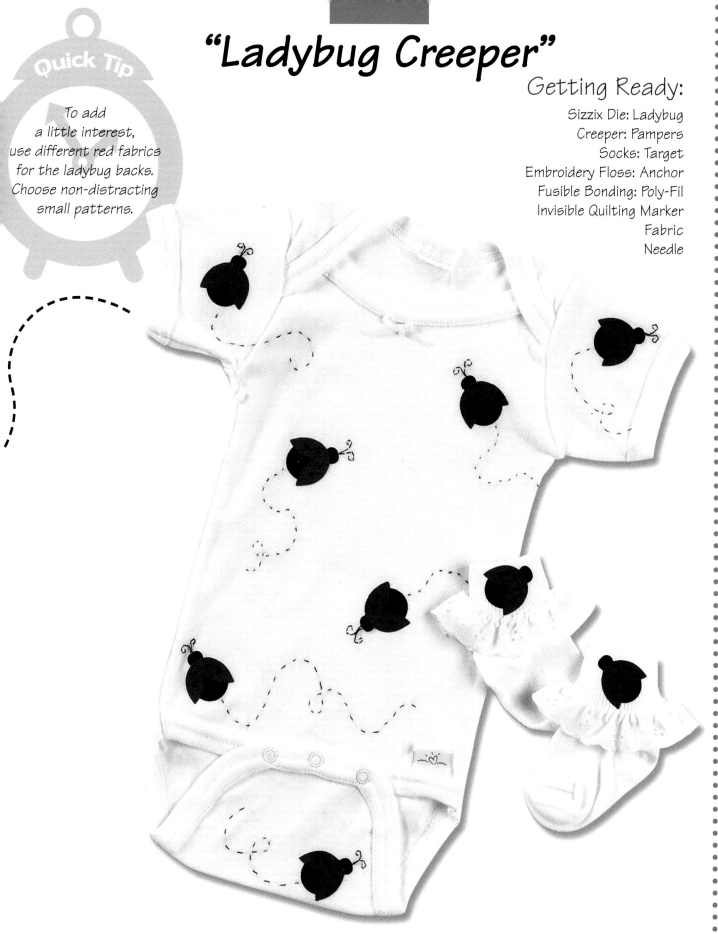

Getting Ready:

Sizzix Die: Ladybug
Creeper: Pampers
Socks: Target
Embroidery Floss: Anchor
Fusible Bonding: Poly-Fil
Invisible Quilting Marker
Fabric
Needle

Technique:

Use an invisible quilting marker to draw embellishments that will be stitched.
The pen marks will disappear after about an hour.

"Accessorize,
Poly Foam Necklaces

Getting Ready:

Sizzix Dies: Filmstrip (Squares), Squares,
Shadow Box Alphabet
Poly Foam: Ellison
Adhesive: Xyron or Hold the Foam glue
Ribbon

Technique:

Sandwich narrow ribbon between matching die-cuts for necklaces that are sure to bring hugs and kisses from your kids.

Accessorize"

Poly Foam Pins & Bolo

Getting Ready:

Sizzix Dies: Cowboy Hat, Mittens, Watermelon
Poly Foam: Ellison
Wire: Artistic Wire
Self-Adhesive Pin Backs: Shrocks
Adhesive: Xyron or Hold the Foam glue
Large Paper Clip
Ribbon

Technique:

Self-adhesive pin backs turn embellished foam die-cuts into cool kids jewelry. Even the boys can be included with a bolo tie.

"Key Chain"

Getting Ready:

Sizzix Die: Flower (Daisy #2)
Shrink Film: Ellison
Embroidering Thread
Punch: Fiskars, 1/8" hole punch
Pens: Sakura Permapaque

Quick Tip

If you don't have permanent marking pens, use colored pencils. The plastic must be sanded first with fine grade sand paper.

Technique:

Die-cut Shrink Film and color with permanent pens. Punch a hole using a hand punch before shrinking to provide a spot to hang the finished shape. Remember the hole shrinks too, so use a punch that is larger than the embroidery floss.

"Flower Bracelet"

Getting Ready:

Sizzix Dies: Flower (Daisy #2), Leaf #1
Shrink Film: Ellison
Colored Pencils: Prismacolor
Permanent Marker: Sharpie
Punch: Fiskars, 1/8" hole punch
Embossing Heat Tool: Marvy/Uchida
Beading Elastic
Beads
Jump Rings
Nail

Quick Tip

Use colored pencils on the rough side of the Shrink Film and permanent markers on the smooth side.

Technique:

Shrink Film can be shrunk in a toaster oven (according to directions) or with an Embossing Heat Tool. Use a nail to hold the plastic where it has been hole-punched. Wave the Embossing Heat Tool slowly and evenly over the Shrink Film.

Cut-It Make-It Wear-It

"Fun Flip Flops"

Getting Ready:

Sizzix Dies: Bat, Hearts, Flower (Daisy #2), Ladybug
Poly Foam: Ellison
Punches: Fiskars, 1/16" & 1/8" hole punches
Wire: Artistic Wire
Sandals: Fresh Produce

Technique:

Punch holes in the top layer of Poly Foam for the design. Thread matching colored wire through the top layer and through the bottom layer before fastening to sandal straps. No adhesive is necessary, as the wire will hold the two layers together while also providing the length to wind around the sandal strap.

Cut-It Make-It Wear-It

70

"More Flip Flops"

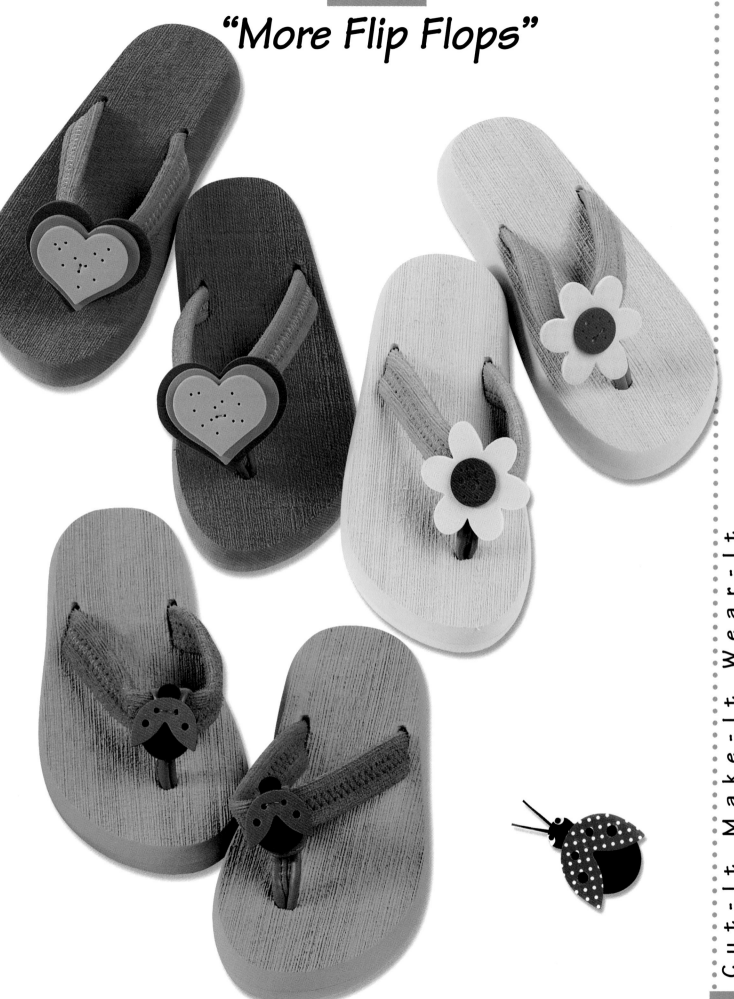

"Pet Tags"

Getting Ready:

Sizzix Dies: Dog, Tags
Shrink Film: Ellison
Colored Pencils: Prismacolor
Permanent Pens: Sharpie
Punch: Fiskars, 1/8" hole punch
Pen: Sakura
Oven or Toaster Oven
Ribbon

Quick Tip

Shrink Film shrinks to less than half it's original size, so make sure to write large enough to read it once it's shrunk.

Technique:

Because Shrink Film is clear, it is easy to print out the text on the computer and trace it onto the film before shrinking. Because the drawing starts out so much larger, it is a great project for kids.

"Flower Lei"

Getting Ready:

Sizzix Die: Flower (Daisy #2)
Plain Cardstock: Provo Craft
Patterned Cardstock: Provo Craft
Punch: Marvy/Uchida, 1/8" hole punch
Cord: Darice
Adhesive: Xyron

Technique:

Die-cut multiple small and large Daisies in several plain and patterned cardstock colors that have been Xyroned. Attach like Daisies back-to-back, then hole-punch. Knot the cord between Daisies to keep them separate.

"Forest Apron"

Getting Ready:

Sizzix Dies: Moon, Pine Trees, Primitive Stars,
Road & Wavy Border, Fun Serif Alphabet
Fusible Bonding: Poly-Fil
Apron
Fabric
Scissors

Quick Tip

Fabric can be put through a Xyron machine as an alternative to Fusible Bonding if laundering isn't required.

Technique:

Die-cut fabric into fun shapes. Use non-distracting patterns in similar color families.
Layout the design before adhering so the shapes can be manipulated into a pleasing design.

"T-Shirts"

Getting Ready:

Sizzix Dies: Flower (Daisy #2), Stars
Iron-on Adhesive: Steam-A-Seam
T-Shirts
Fabrics
Button
Thread

Quick Tip

Add a scrap of paper as an extra layer when die-cutting fabric. This helps to eliminate "shaggy" edges on the fabric die-cut.

Technique:

Stitch a rectangle of fabric onto the T-shirt first, if the T-shirt color doesn't coordinate with the fabric die-cuts. Remove one side of backing and attach iron-on adhesive to fabric BEFORE cutting. This allows cutting the fabric and adhesive together. After cutting the shapes, remove the remaining adhesive backing and position fabric die-cuts on the shirt before ironing with a press cloth.

Home & Play

Personalize your home décor with Sizzix. Whether cutting paper, plastic, cork or fabric the result is the same... Fantastic! Create cool kids games and personalize your party decorations with just a little Sizzix Machine and a lot of imagination.

"Fancy Frames"

Getting Ready:

Sizzix Dies: Hearts, Rectangle Frame, Scallop Frame
Plain Cardstock: Provo Craft & Sizzix
Patterned Paper: Provo Craft
Micro-Beads: Art Accents
Tape Sheets: Art Accents
Eyelets: Impress
Magnet: Xyron

Technique:

Adhere the tape sheets to a piece of white cardstock before die-cutting. Pour the beads into a tray big enough to hold the Rectangle Frame. Expose the top side of the tape and lay in the beads.

Home & Play

77

"Drink Tags"

Getting Ready:

Sizzix Dies: Balloons #1, Balloons #2,
Flower (Daisy #2), Snowman
Punch: Fiskars, 1/4" hole punch
Shrink Film: Ellison
Pens: Sakura
Drinking Glasses with Stems
Ball Chain
Baking Pan
Oven

Quick Tip

Drink tags
aren't just for drinks.
Many desserts fit
perfectly into glass or
plastic containers
with stems.

Technique:

Punch holes in each tag before shrinking. Test the shrunken tag to be sure the ball chain will fit through the hole. If ball chain doesn't fit, punch two or three holes in a cluster to make a larger opening (before shrinking). Holes can also be drilled after shrinking with a drill bit that is slightly larger than ball chain.

Technique:

Design the detail for each tag on paper first. After die-cutting the Shrink Film, place the plastic shape over the paper, trace and color the design with permanent pens. Print or handwrite the name last (over the colored pen).

"Easter Table Runner"

Getting Ready:

Sizzix Dies: Basket, Bow, Bunny, Eggs, Jelly Beans
Fusible Bonding: Poly-Fil
Sewing Machine
Scissors
Fabrics
Batting
Needle

Quick Tip

Stitch "hopping" lines behind the bunnies to create a border.

Technique:

Create dimension with the grass in this fabric project. Cut thin strips of fabric with fusible bonding already attached. Use just the tip of the iron to press the edges of the strips, leaving loops between pressed areas for a lumpy, grassy look.

"Turkey Tracks Table Runner"

Getting Ready:

Sizzix Dies: Leaf #1, Turkey
Fusible Bonding: Poly-Fil
Embroidery Floss: Anchor
Quilters Invisible Marker
Sewing Machine
Scissors
Fabric

Quick Tip

This project also makes a great holiday wall hanging.

Technique:

Use small scraps of fabric for each feather in the Turkey's tail. Choose Autumn inspired hues like golds, nutmegs, reds and browns.

"Checkers"

Getting Ready:

Sizzix Dies: Circles, Squares, Fun Serif Alphabet
Stiffened Felt: CPE Inc.
Glue: Elmer's
Cardboard

Quick Tip

Title your game board on the back side with die-cut letters.

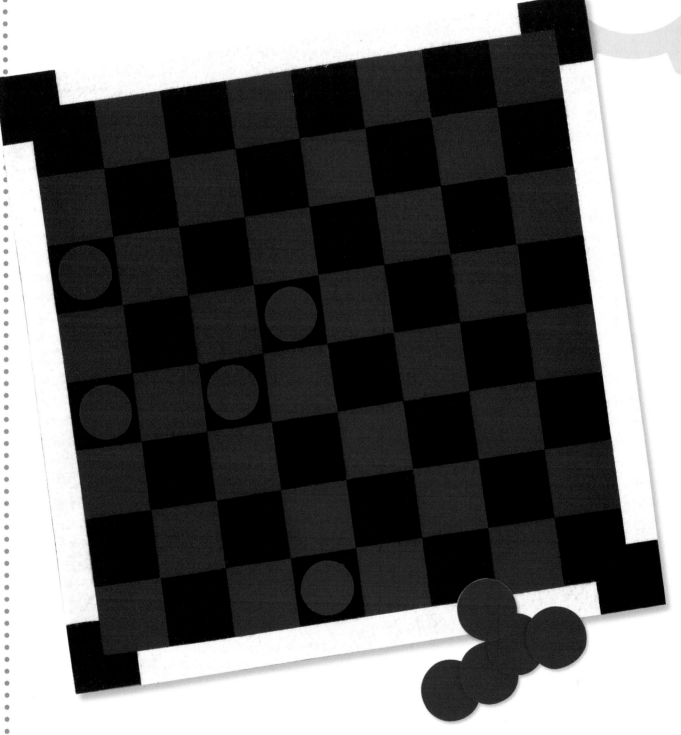

Technique:

Craft a fun game board by die-cutting felt squares and adhering them to cardboard.
Finish off the board with a white border and black corners. Die-cut Circles for game pieces.

"Finger Puppets"

Getting Ready:

Sizzix Dies: Bird, Dolphin, Frog, Grass, Pig, Sea Horse, Wave
Ellison Dies: Plain Finger Puppet
Plain Paper: Tru-Ray & Canson
Patterned Paper: Paper Patch
Punches: Fiskars, 1/8" & 1/16" hole punches
Pop Dots: All Night Media
Adhesive: Xyron
Mini-stapler (optional)

Quick Tip

Attach each puppet to the Finger Puppet base with Pop Dots to make the animals really stand out from the base and for an extra strong hold.

Technique:

Decorating die-cuts is easy. Cut the same shape from multiple colors. Cut out features from one color and attach to the base shape OR punch out designs from the top layer and place solid shape underneath allowing the designs to show through. Some die-cuts have features that fall out in the cutting process (i.e. bird's beak). Simply fasten a small piece of contrasting colored paper behind the die-cut opening for a yellow bird with an orange beak.

Home & Play

83

"Magnets"

Getting Ready:

Sizzix Dies: Angel, Palm Tree, Snowman, Train
Plain Paper: Canson, Paper Adventures & Tru-Ray
Patterned Paper: Paper Patch & Provo Craft
Self-Adhesive Magnet: Ellison
Pen: Marvy/Uchida
Adhesive: Xyron
Scissors

Quick Tip

Cut each shape in multiple colors of paper. Create depth by cutting away portions of the lighter shade where shadows or folds might be and layering on top of a darker shade.

Technique:

Peel off the paper backing on self-adhesive magnet and attach the base color of paper.
Die-cut with the paper and magnet together then layer paper on top for magnet detail.
This avoids the frustration of trying to line up the finished project on top of the magnet later.

"Holiday Ornaments"

Getting Ready:

Sizzix Die: Santa Head
Plain Paper: Paper Adventures
Patterned Paper: Paper Patch
Colored Pencils: Sanford Prismacolor
Foam Adhesive: All Night Media
Adhesive: Xyron, double stick tape
Scissors
Ribbon

Quick Tip

Add jingle bells to holiday projects for crafts that look and sound great.

Technique:

When projects call for the back to look exactly like the front, create mirror images by cutting the front shape with the adhesive side of the paper down and the back shape with the adhesive side up. The two adhesive sides should face each other, with the ribbon sandwiched in the middle when assembling the ornaments.

"Cork Pine Tree Frame"

Getting Ready:

Sizzix Dies: Photo Corners, Pine Trees
Patterned Paper: Provo Craft
Cork: Craft-T Paper Products
Self-Adhesive Magnet: Ellison
Adhesive: Xyron
Hot Glue Gun & Glue
Twine

Quick Tip

Place magnetic strips behind cork to make this great frame a refrigerator magnet.

Technique:

Use cork and twine to create a natural, outdoorsy look to a photo. Knot the end of three strands of twine and braid. Hide the ends under a die-cut Pine Tree.

"Snowman Pillow"

Getting Ready:

Sizzix Dies: Branch & Leaves, Campfire, Squares
Fusible Bonding: Poly-Fil
Pens: Sakura
Sewing Machine
Scissors
Buttons
Batting
Fabric
Thread

Technique:

Squares in various sizes can be used to create a whimsical snowman. Embellish with buttons and scraps of fabric in coordinating colors and patterns.

"Patterns"

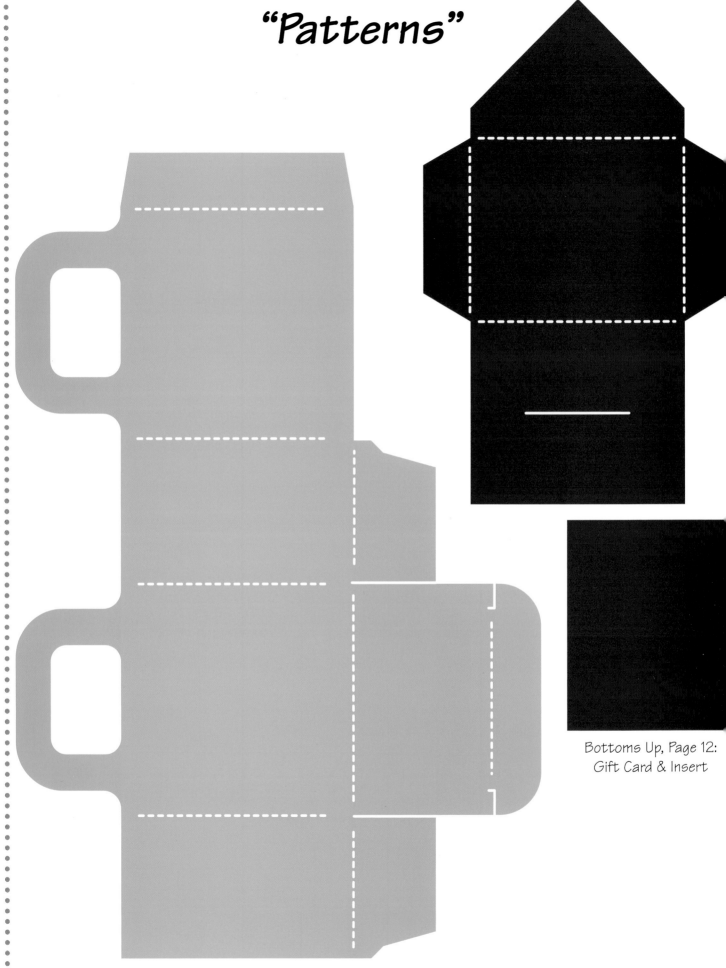

Bottoms Up, Page 12:
Gift Card & Insert

Decorative Boxes, Page 56: Bag #1

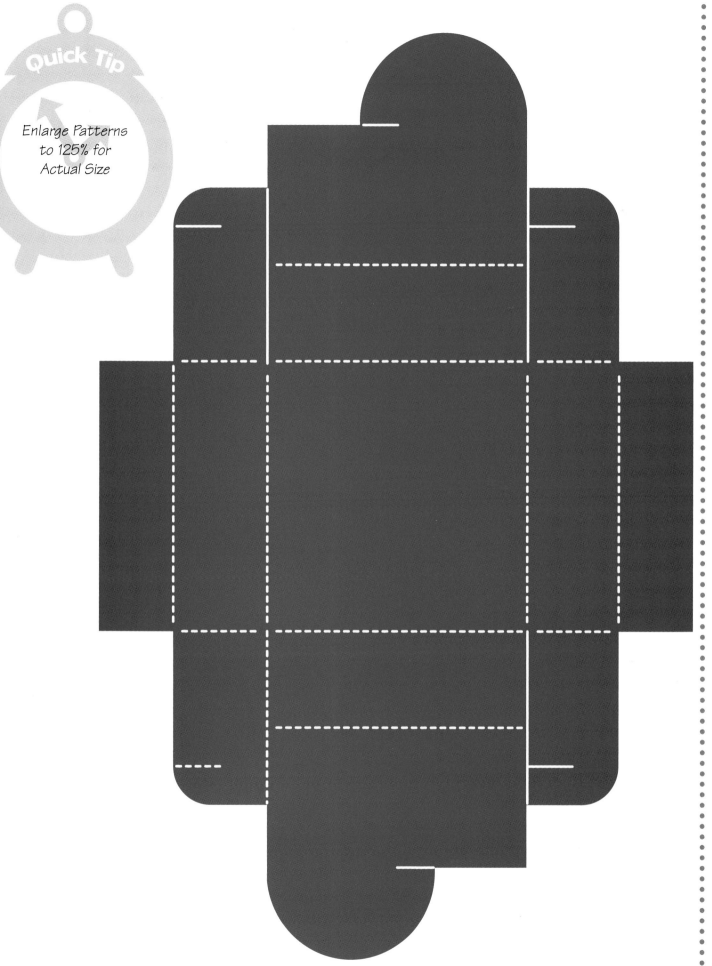

Critter Boxes, Page 61: Box #2

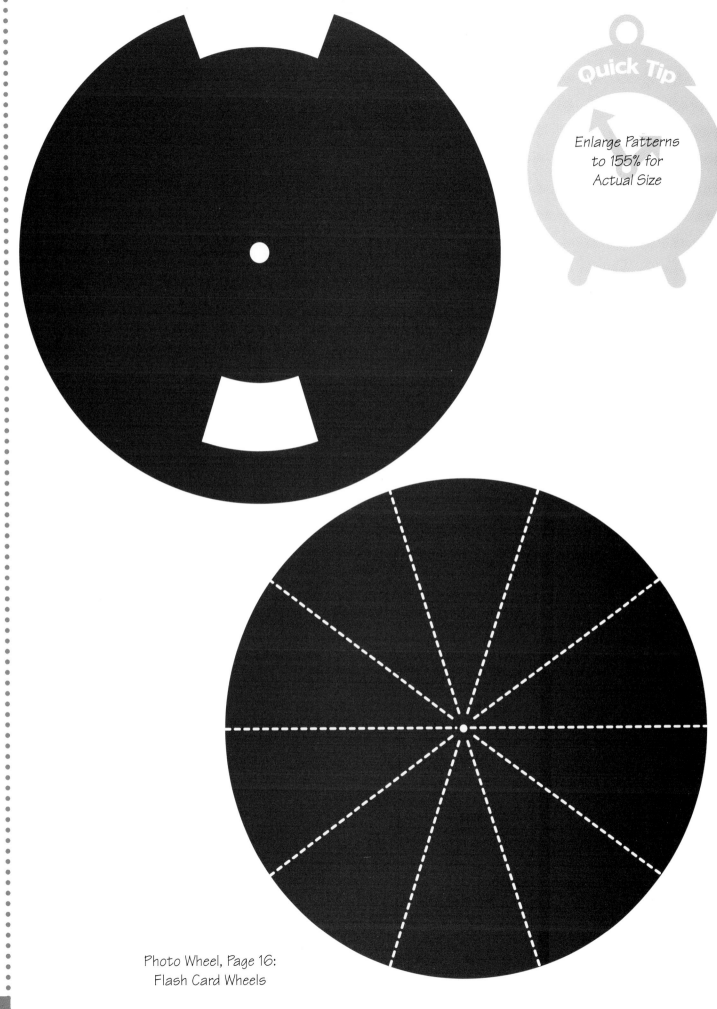

Quick Tip

Enlarge Patterns
to 155% for
Actual Size

Photo Wheel, Page 16:
Flash Card Wheels

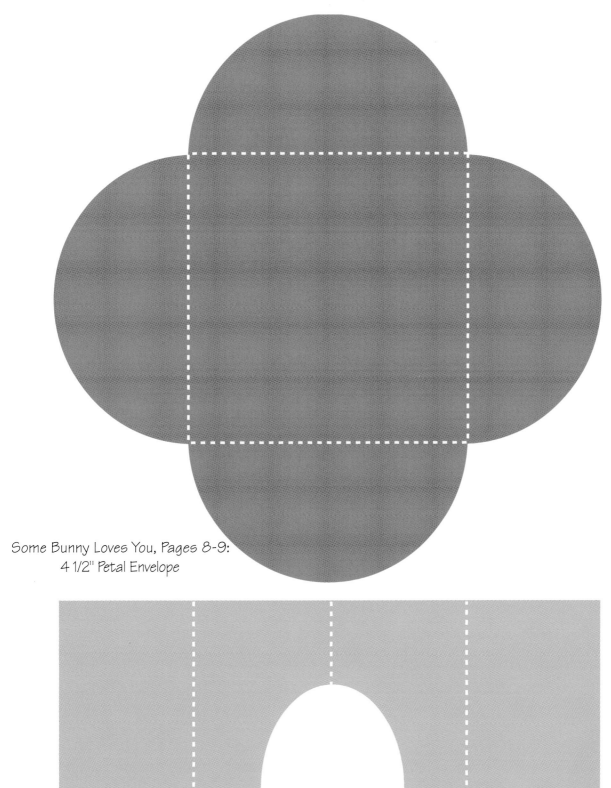

Some Bunny Loves You, Pages 8-9:
4 1/2" Petal Envelope

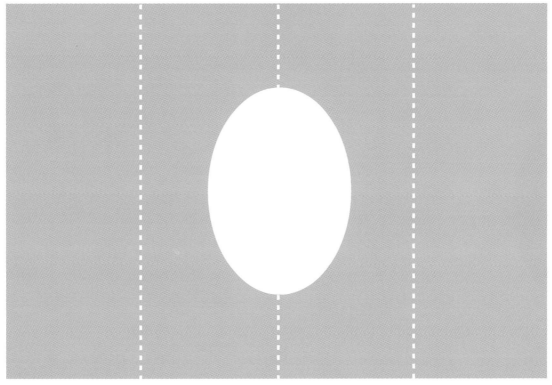

Dragonfly Dangle, Page 47:
Accordion w/Cut-Out

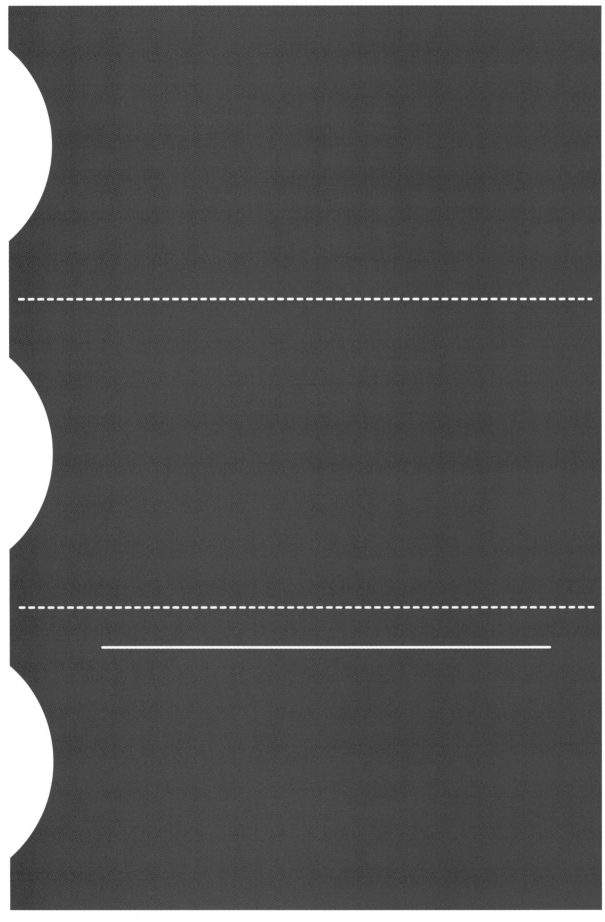

Flower Slider, Pages 34-35: Slider Card Set

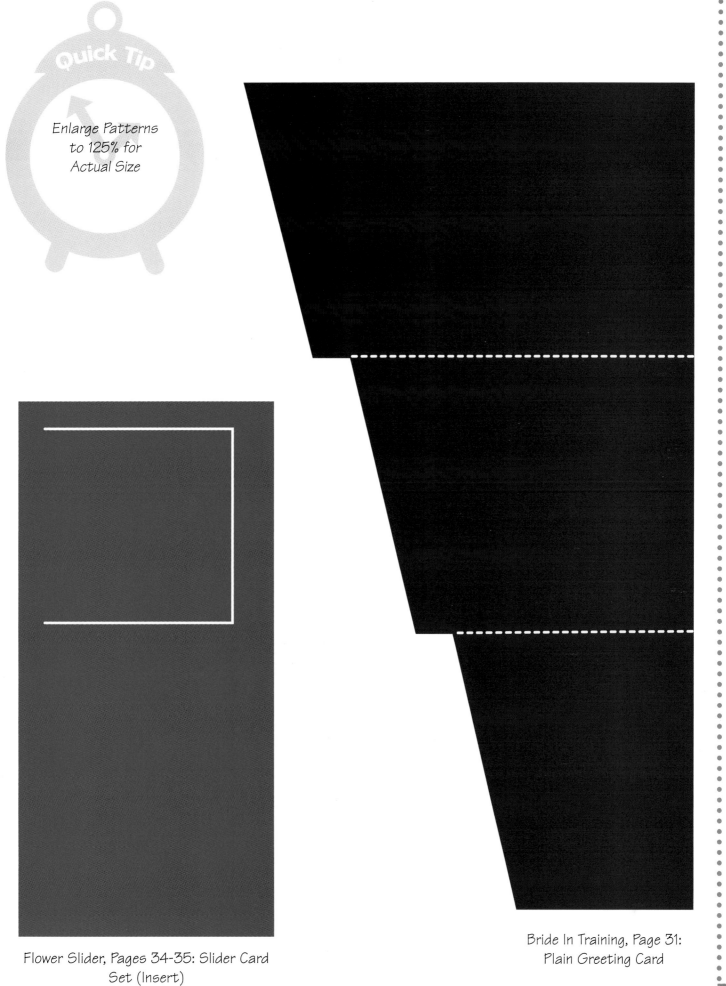

Flower Slider, Pages 34-35: Slider Card
Set (Insert)

Bride In Training, Page 31:
Plain Greeting Card

Quick Tip

Enlarge Patterns
to 125% for
Actual Size

Happy New Year, Pages 44-45:
Photo Envelope #1

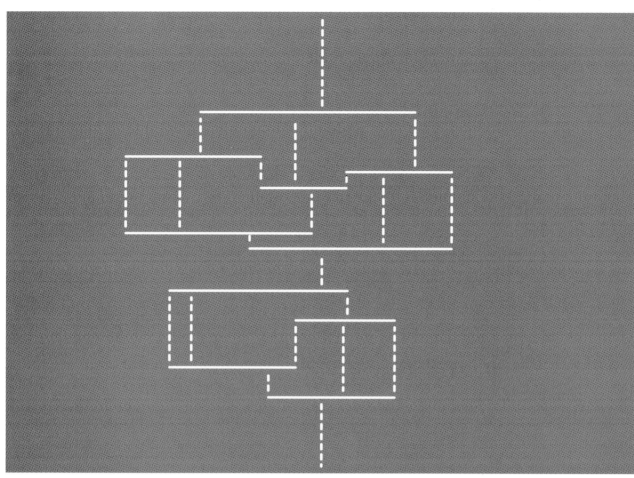

I Be "Leaf" In You, Page 50: Multiple Pop-Up Card #1

Happy Halloween, Page 51: Diorama Arch Card

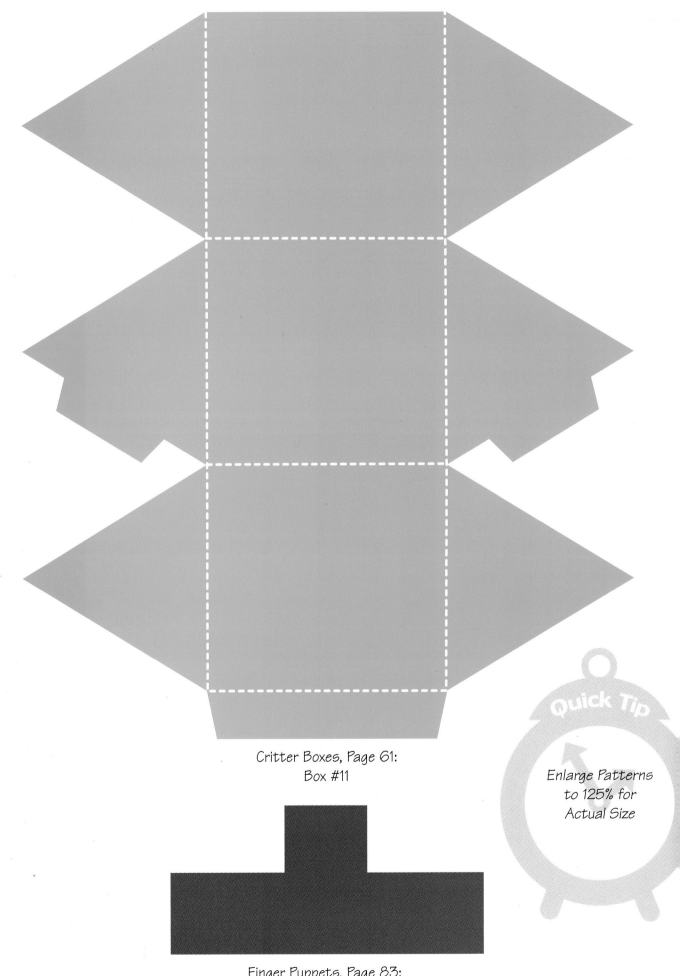

Critter Boxes, Page 61:
Box #11

Quick Tip

Enlarge Patterns
to 125% for
Actual Size

Finger Puppets, Page 83:
Plain Finger Puppet